3-4歲

幼兒全方位
智能開發

英文篇

Let's Learn ABC
大楷字母

園丁文化

Capital Letter A
大楷字母 A

做得好！ 不錯啊！ 仍需加油！

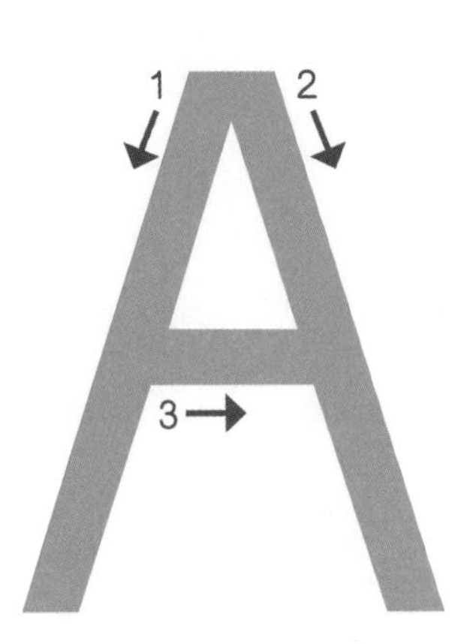

Apple 蘋果

Ant 螞蟻

Circle the letter "A"s. 請圈出大楷字母 A。

Let's write the letter A. 齊來寫一寫。 

A	A				

Capital Letter B
大楷字母 B

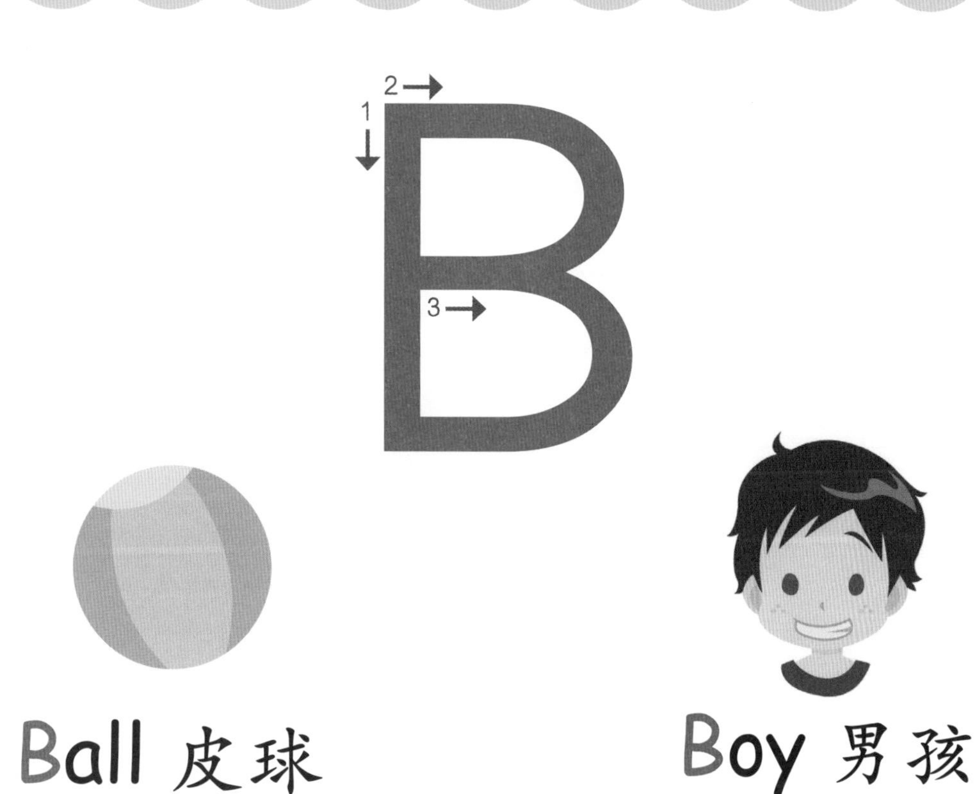

Fill in the letter B. 請填寫大楷字母 B。

Let's write the letter B. 齊來寫一寫。

Capital Letter C
大楷字母 C

做得好！ 不錯啊！ 仍需加油！

Cat 貓　　Car 汽車

Circle the thing that starts with the letter C. 請圈出以字母 C 開頭的物件

Let's write the letter C. 齊來寫一寫。 

C	C				

答案：Cake

Capital Letter D
大楷字母 D

Dog 狗　　Doll 洋娃娃

What is this? Circle the correct word. 這是什麼？請圈出正確的詞彙。

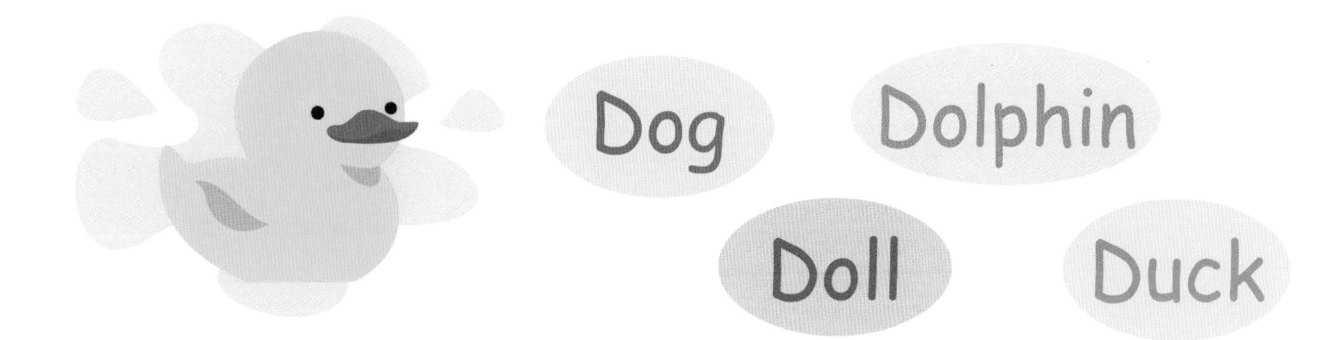

Let's write the letter D. 齊來寫一寫。

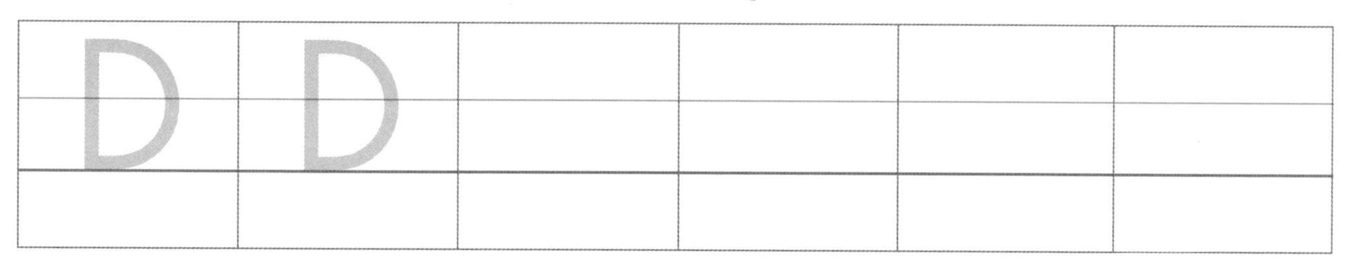

答案：Duck

Capital Letter E
大楷字母 E

Elephant 象

Egg 雞蛋

Colour the Easter eggs. 請把復活蛋填上顏色。

Let's write the letter E. 齊來寫一寫。 

E	E				

Let's review A to E
温習 A 至 E

做得好！ 不錯啊！ 仍需加油！

- Match each letter with the correct word and picture.
請把字母和正確的詞彙及圖畫用線連起來。

	Letter	Word		Picture
1.	A	Boy	A.	
2.	B	Doll	B.	
3.	C	Apple	C.	
4.	D	Elephant	D.	
5.	E	Cat	E.	

答案：1. Apple, B　2. Boy, E　3. Cat, A　4. D, Doll, D　5. Elephant, C

Capital Letter F
大楷字母 F

做得好！ 不錯啊！ 仍需加油！

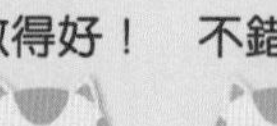

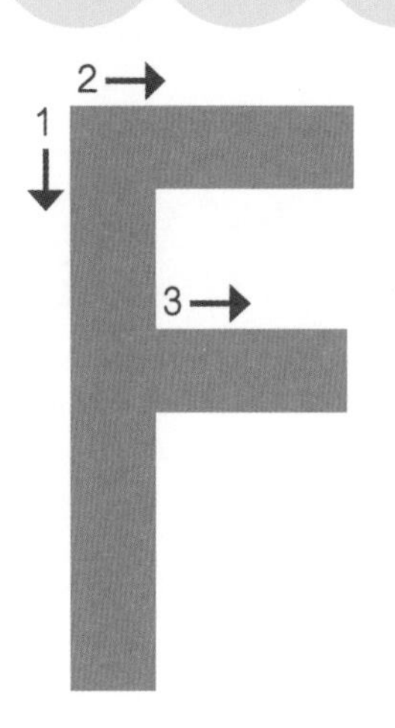

Flower 花　　　　Fish 魚

Fill in the letter F. 請填寫大楷字母 F。

Let's write the letter F. 齊來寫一寫。

F	F				

Capital Letter G
大楷字母 G

Girl 女孩　　Glass 玻璃杯

- Circle the letter “G” s . 請圈出大楷字母 G。

- Let's write the letter G. 齊來寫一寫。

G	G				

Capital Letter H
大楷字母 H

Hat 帽子　　Hen 母雞

Draw a line to match the picture with the correct word.
請把圖畫和正確的詞彙用線連起來。

1. 

- Horse
- House

2. 

Let's write the letter H. 齊來寫一寫。 

H	H				

答案：1. House　2. Horse

Capital Letter I
大楷字母 I

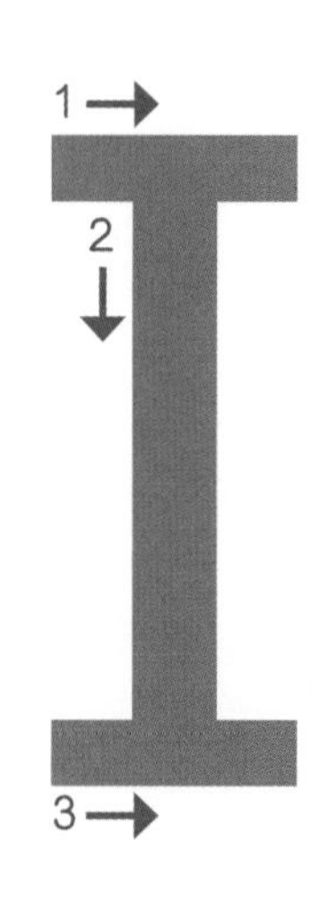

Ice 冰

Insect 昆蟲

Help the boy find the ice cream by colouring the letter "I"s.
請把大楷字母 I 填上顏色，幫助小男孩找到雪糕吧。

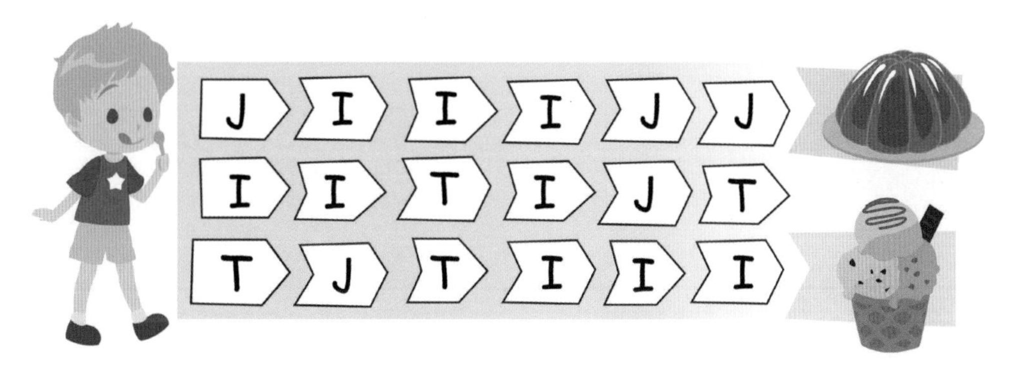

Let's write the letter I. 齊來寫一寫。 

I	I				

Capital Letter J
大楷字母 J

做得好！ 不錯啊！ 仍需加油！

  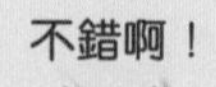  

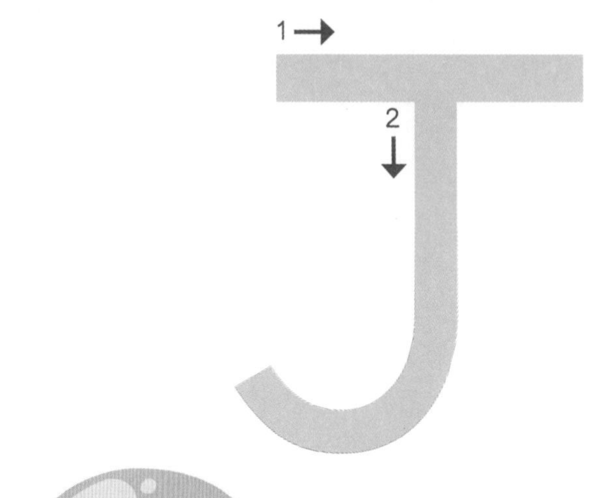

Jellyfish 水母

Jacket 外套

What is this? Circle the correct word. 這是什麼？請圈出正確的詞彙。

Jacket Jet
Juice Jam

Let's write the letter J. 齊來寫一寫。 

J	J				

答案：Juice

Let's review F to J
温習 F 至 J

- Circle the letters F, G, H, I and J in the tasty bowl of alphabet soup below.
  這鍋美味的湯裏有很多字母意粉，請把大楷字母 F、G、H、I 和 J 圈起來。

Capital Letter K
大楷字母 K

做得好！ 不錯啊！ 仍需加油！

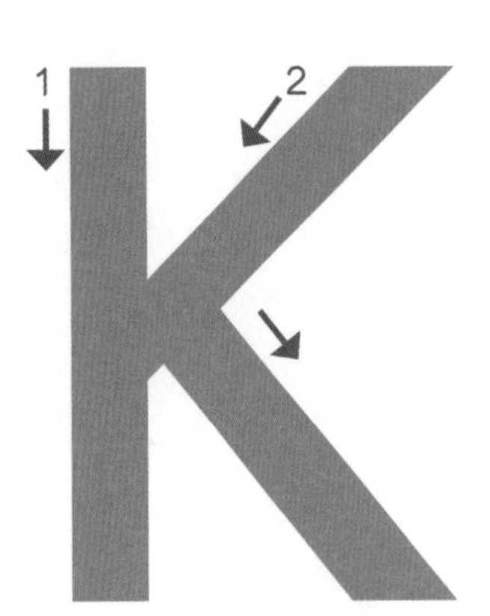

Kite 風箏　　Kangaroo 袋鼠

Colour the kite with the letters that spell "KITE".
請把寫上 KITE 的風箏填上顏色。

1. 

2. 

3. 

Let's write the letter K. 齊來寫一寫。 

K	K				

答案：2

Capital Letter L
大楷字母 L

Leaf 葉子

Lion 獅子

● Circle the thing that starts with the letter L. 請圈出以字母 L 開頭的物件。

● Let's write the letter L. 齊來寫一寫。 

L	L				

答案：Lollipops

Capital Letter M
大楷字母 M

做得好！ 不錯啊！ 仍需加油！

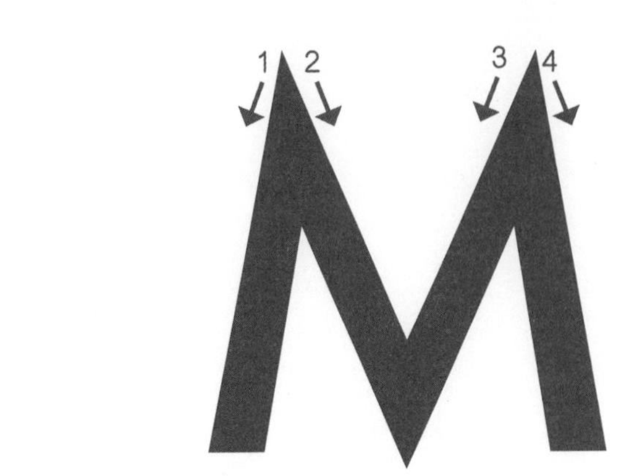

Monkey 猴子

● Circle the letter "M"s . 請圈出大楷字母 M。

● Let's write the letter M. 齊來寫一寫。 

Capital Letter N
大楷字母 N

做得好！  不錯啊！  仍需加油！ 

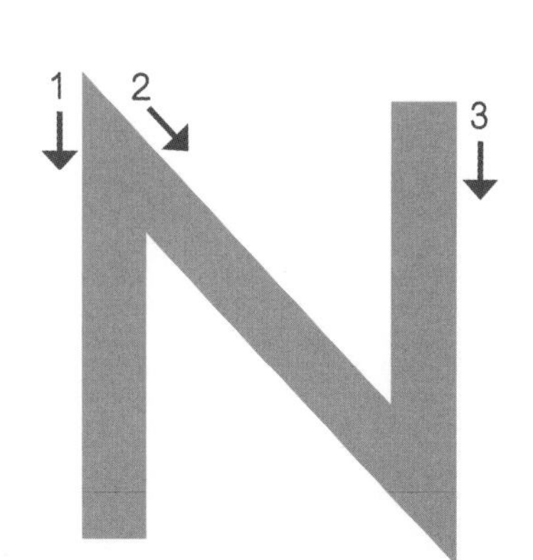

Nest 鳥巢

Draw a nose for the pictures below. 請給下面的人物或動物畫上鼻子。

Let's write the letter N. 齊來寫一寫。 

N	N				

Capital Letter O
大楷字母 O

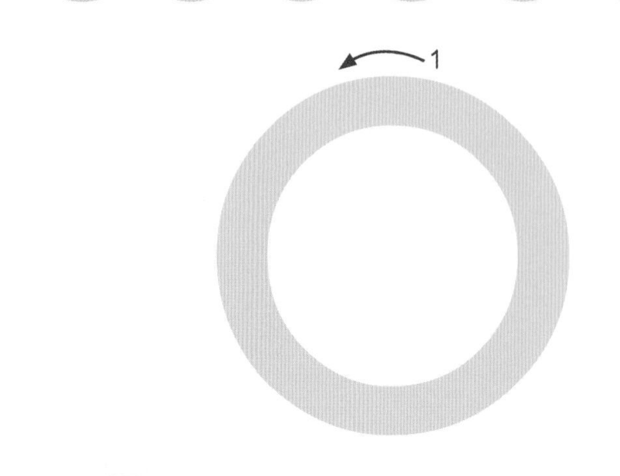

Orange 橙　　Owl 貓頭鷹

Colour the fruit below as indicted. 請把水果填上正確顏色。

Let's write the letter O. 齊來寫一寫。 

O	O				

Let's review K to O
温習 K 至 O

Welcome to the zoo! Fill in the missing letter to complete the words.
歡迎來到動物園！請把正確的大楷英文字母填在橫線上。

K L M N O

答案：1. Nest, Owl, Kangaroo, Lion, Monkey

Capital Letter P
大楷字母 P

做得好！ 不錯啊！ 仍需加油！

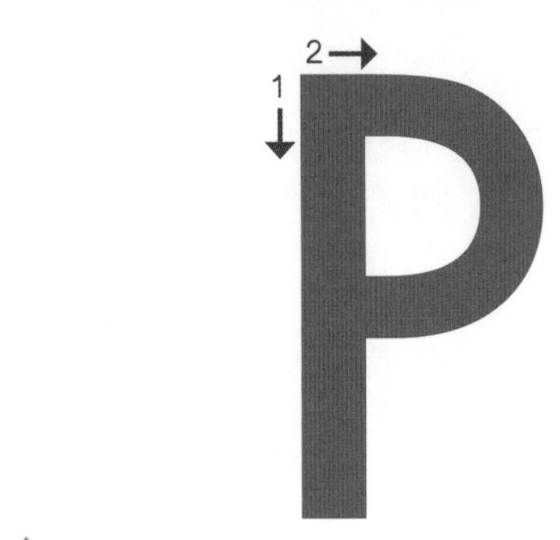

Pig 豬

Pencil 鉛筆

Draw a line to match the picture with the correct word.
請把圖畫和正確的詞彙用線連起來。

1.

Pan

Parrot

2.

Let's write the letter P. 齊來寫一寫。

P	P				

答案：1. Pan 2. Parrot

Capital Letter Q
大楷字母 Q

Circle the two quilts that are the same. 請把兩張相同的棉被圈起來。

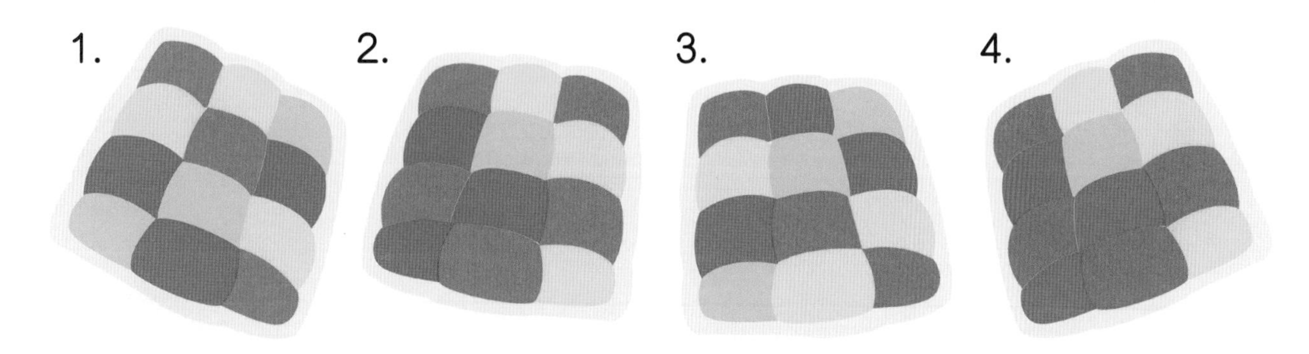

Let's write the letter Q. 齊來寫一寫。

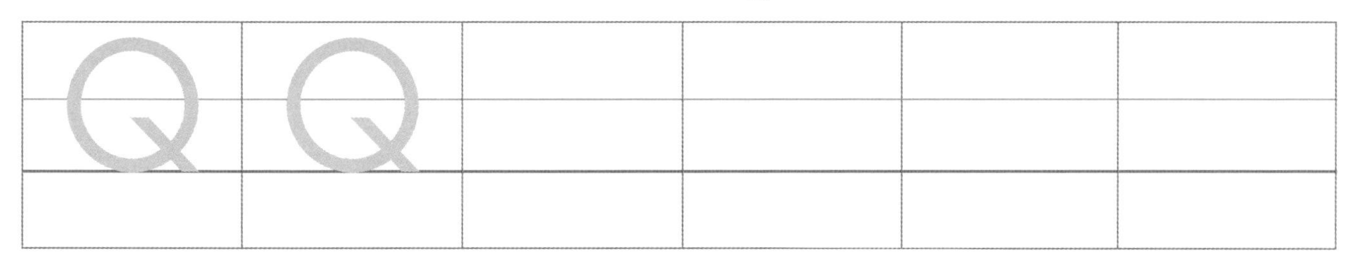

答案：2, 4.

Capital Letter R
大楷字母 R

做得好！  不錯啊！  仍需加油！ 

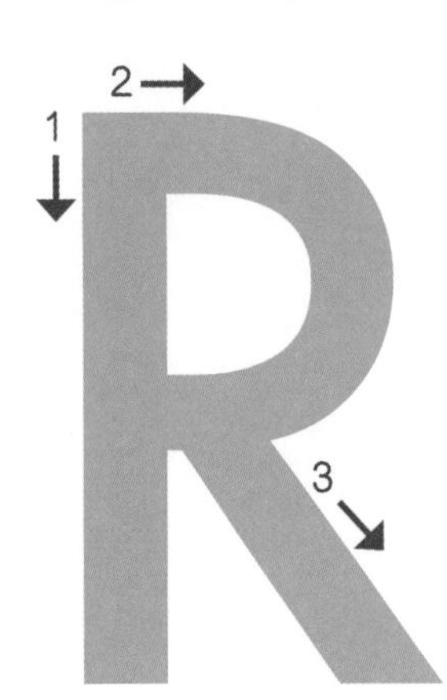

Robot 機械人　　Rain 下雨

- Help the rabbit find the carrot by colouring the letter “R”s.
  請把字母 R 填上顏色，幫助兔子找到紅蘿蔔吧。

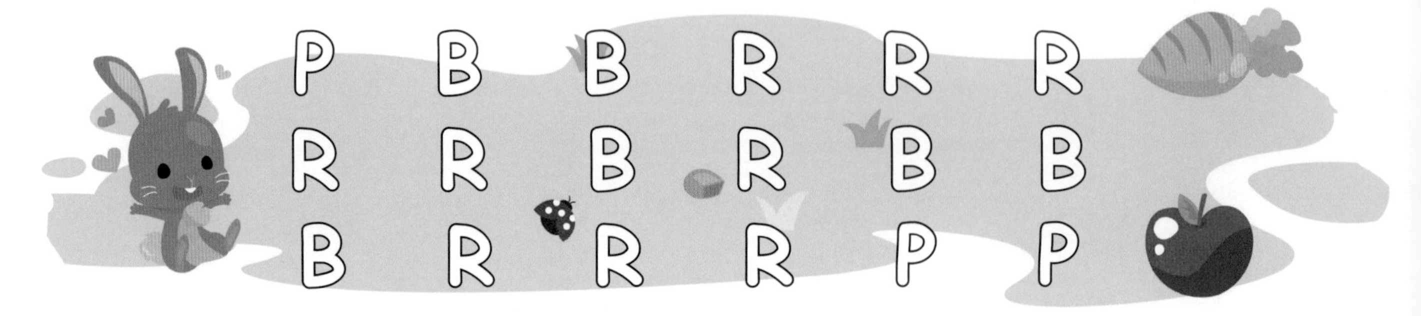

- Let's write the letter R. 齊來寫一寫。 

R	R				

Capital Letter S
大楷字母 S

Sun 太陽

Sheep 綿羊

● What is this? Circle the correct word. 這是什麼？請圈出正確的詞彙。

Sheep Snake
Squirrel Shark

● Let's write the letter S. 齊來寫一寫。 

S	S				

答案：Squirrel

Capital Letter T
大楷字母 T

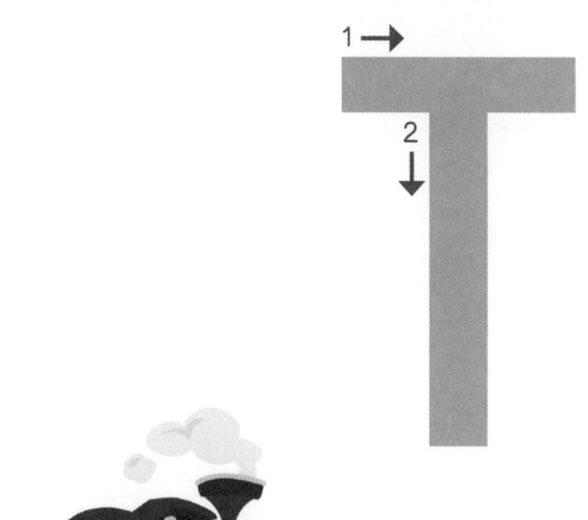

Train 火車

Tiger 老虎

- Circle the thing that starts with the letter T. 請圈出以字母 T 開頭的物件。

- Let's write the letter T. 齊來寫一寫。 

T	T				

答案：Tree

Let's review A to T
温習 A 至 T

- Help the Queen find her crown by following the path from A to T.
  請依 A 至 T 的順序走，幫皇后找到她的皇冠吧。

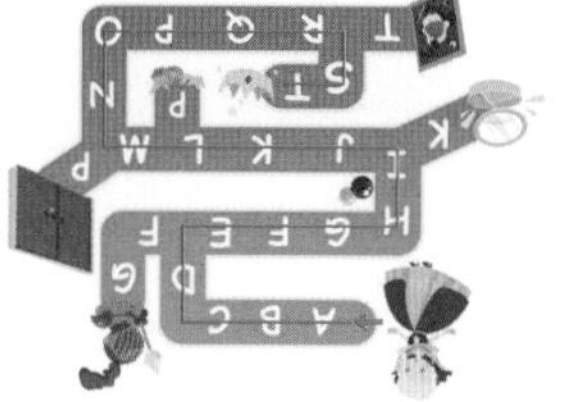

答案：

Capital Letter U
大楷字母 U

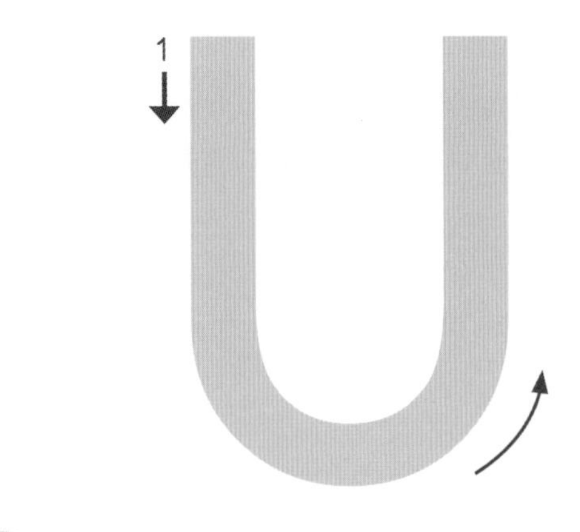

Umbrella 雨傘　　Uniform 校服

Join the same umbrellas together with a line.
請把相同的雨傘用線連起來。

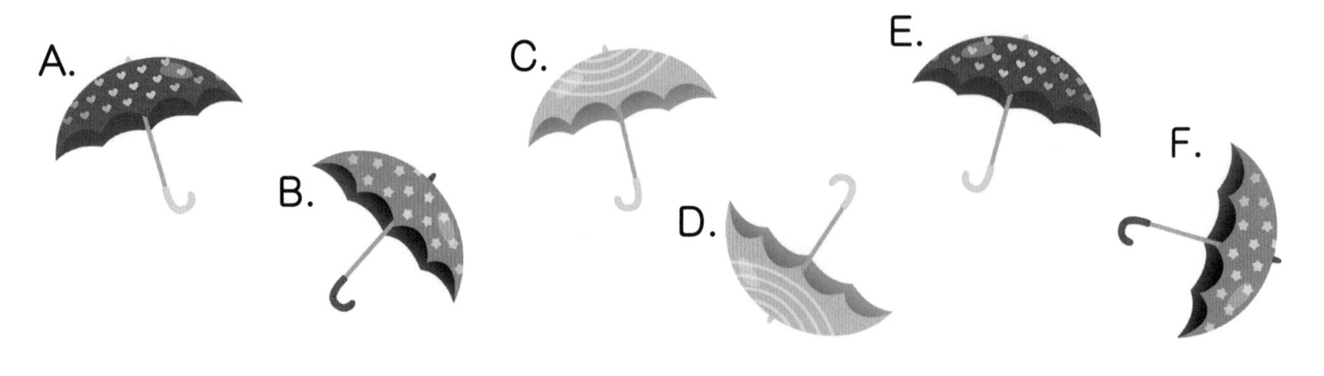

Let's write the letter U. 齊來寫一寫。

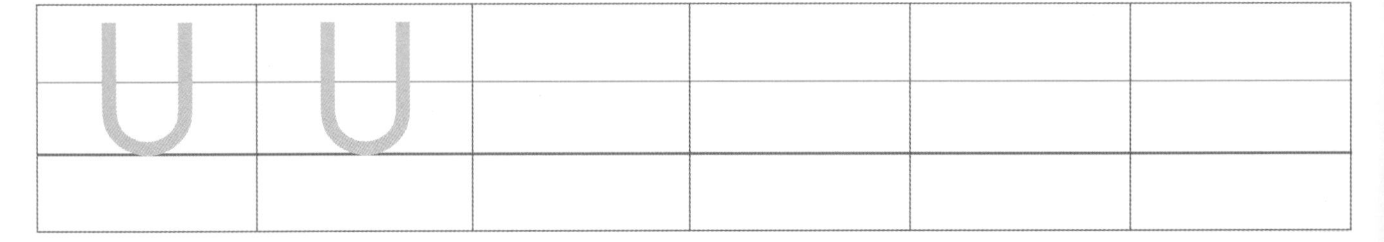

答案：A, E; B, F; C, D

Capital Letter V
大楷字母 V

Violin 小提琴　　　　Vest 背心

- Colour the vest with the letters that spell "VEST".
  請把寫上 VEST 的背心填上顏色。

1.

2.

3.

- Let's write the letter V. 齊來寫一寫。 

V	V				

答案：1

Capital Letter W
大楷字母 W

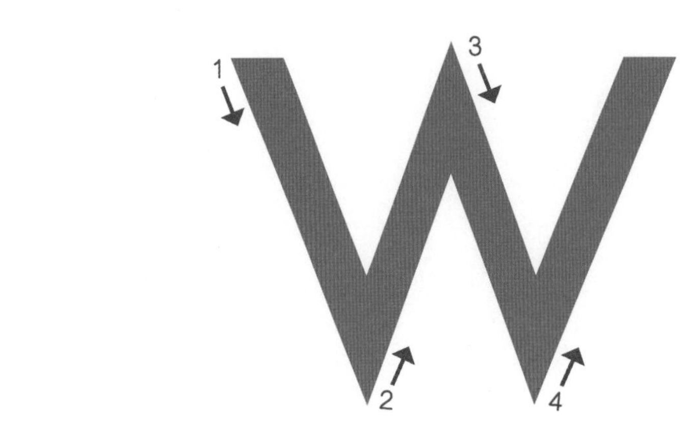

Whale 鯨魚

Water 水

● Fill in the letter W. 請填寫大楷字母 W。

● Let's write the letter W. 齊來寫一寫。 

W	W				

Capital Letter X
大楷字母 X

做得好！ 不錯啊！ 仍需加油！

Xylophone 木琴

X-ray X 光

● Circle the letter “X” s . 請圈出大楷字母 X。

BOX

SIX

FOX

● Let's write the letter X. 齊來寫一寫。 

X	X				

Capital Letter Y
大楷字母 Y

做得好！ 不錯啊！ 仍需加油！

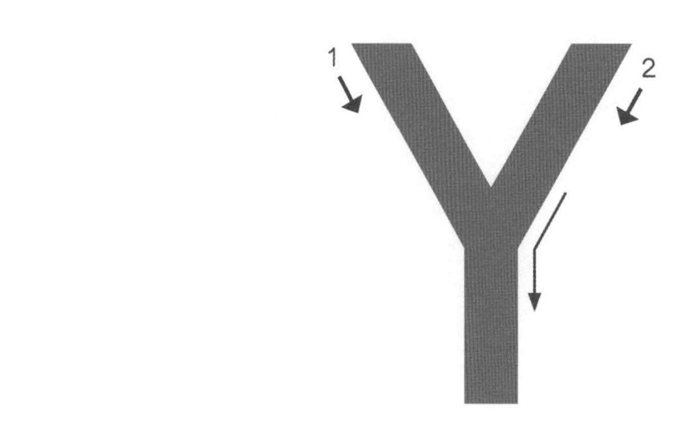

Yacht 遊艇　　Yo-Yo 搖搖

- Colour the yo-yos with the letter Y in yellow.
  請把寫有字母 Y 的搖搖填上黃色。

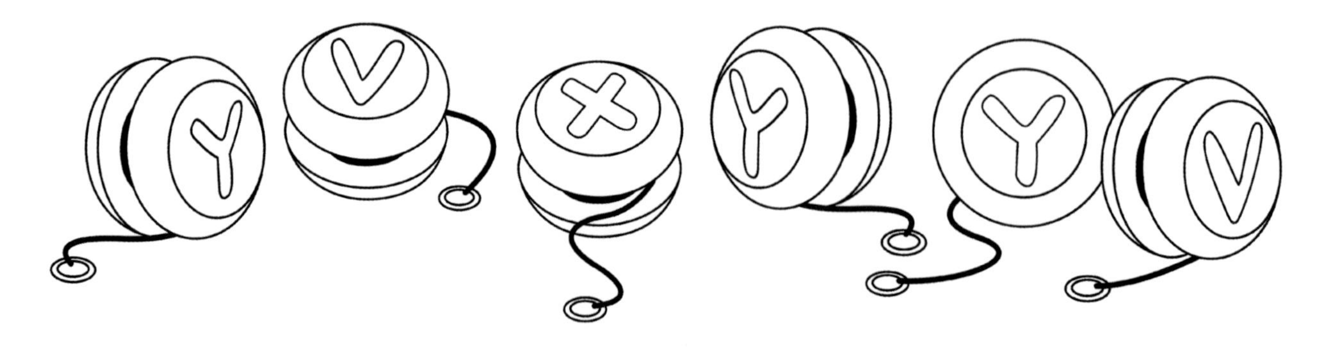

- Let's write the letter Y. 齊來寫一寫。 

Y	Y				

Capital Letter Z
大楷字母 Z

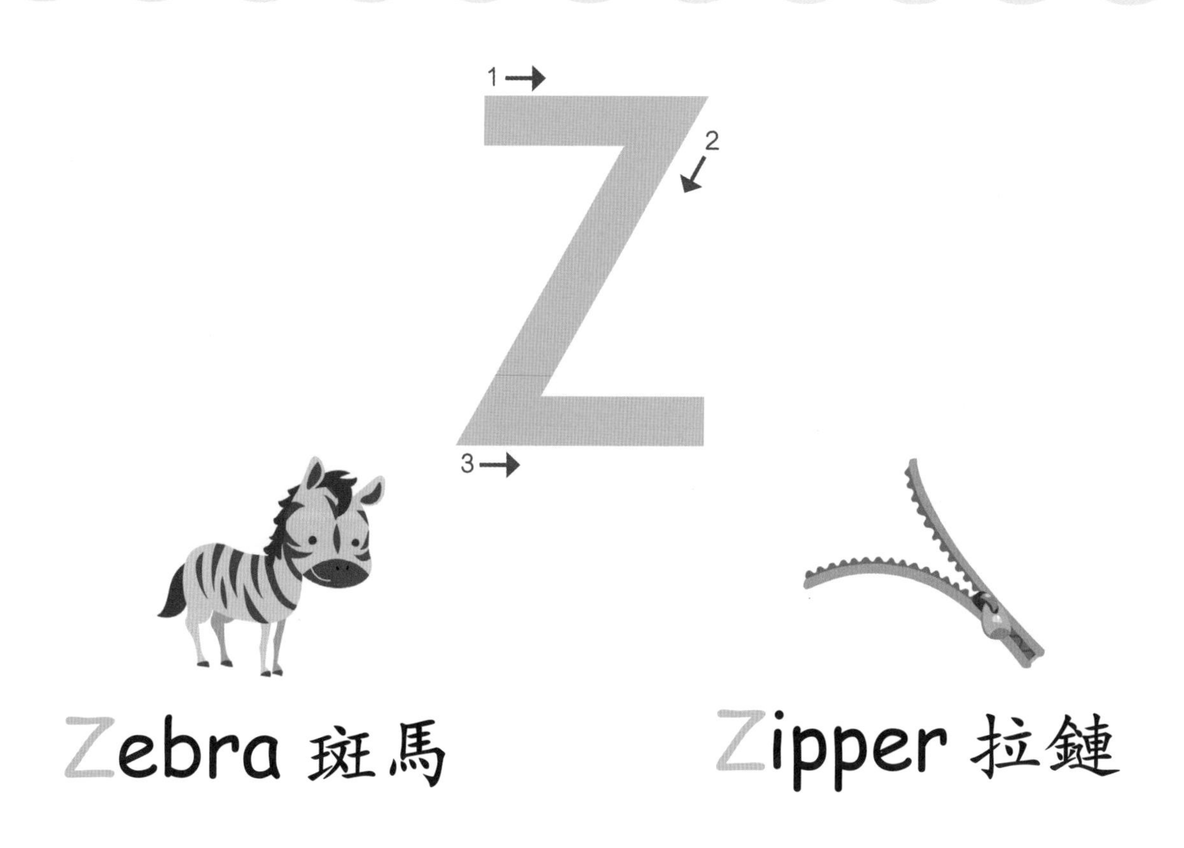

- Circle the thing that starts with the letter Z. 請圈出以字母 Z 開頭的物件。

- Let's write the letter Z. 齊來寫一寫。

Z	Z				

答案：Zoo

Let's review A to Z
温習 A 至 Z

做得好！ 不錯啊！ 仍需加油！

- Fill in the missing letters in alphabetical order.
  請依英文字母的順序，在車卡上填寫缺少了的大楷字母。

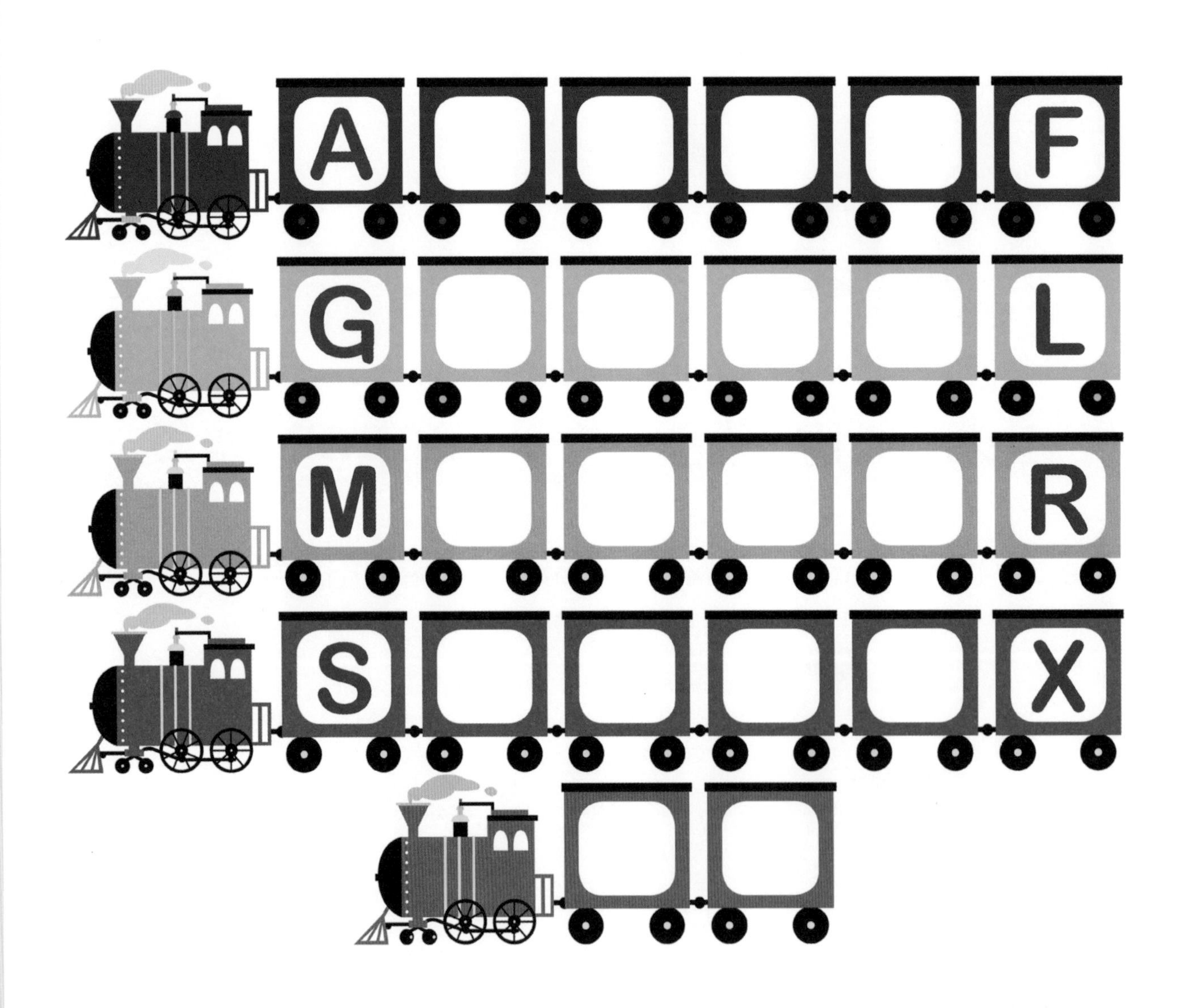